The Restless Earth

Monica Sevilla

Contents

Plate Tectonics
Tectonic Plate Boundaries
Continental Drift
Mid Ocean Ridges and Seafloor Spreading
Subduction
What is a Volcano?
Mt. Vesuvius
The Yellowstone Mantle Plume
What is an Earthquake?
What is a Tsunami?

Plate Tectonics

Plate tectonics is the theory that that explains how the continents move upon the Earth. A **theory** is an educated guess or a prediction of what you think will happen. **Tectonic plates** are sections of the Earth's crust or lithosphere that move upon the mantle of the Earth. The **mantle** is made up of dense,

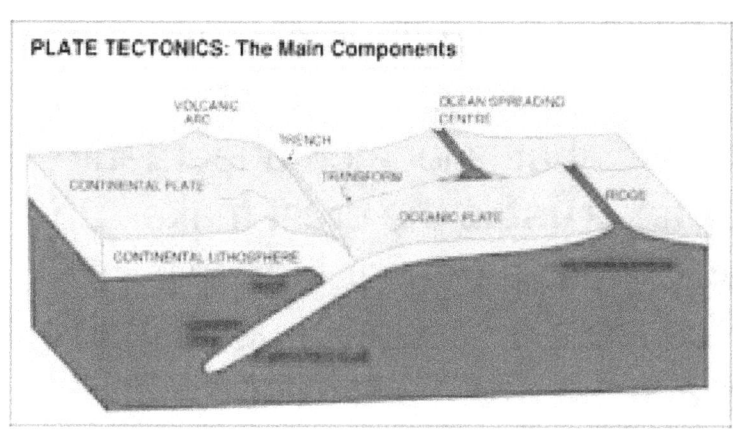

molten or melted rock. The tectonic plates are less dense than the mantle and float on top of it. **Density** is the amount of matter in a given space or volume. The mantle has more matter per unit volume than the crust or lithosphere. This is why the mantle and the crust or lithosphere separates into layers.

The Earth has 9 major tectonic plates. The tectonic plates are known as: the North American plate, Pacific plate, Nazca plate, South American plate, African plate,

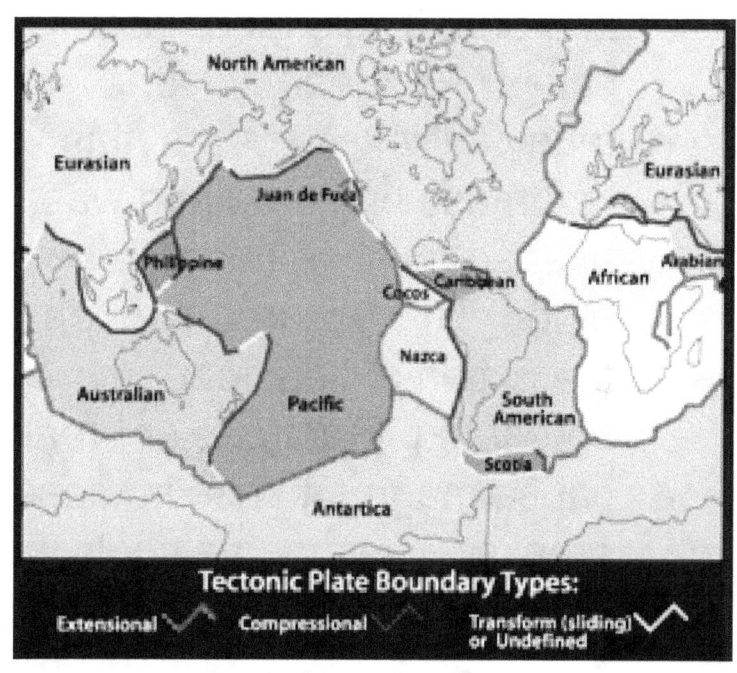

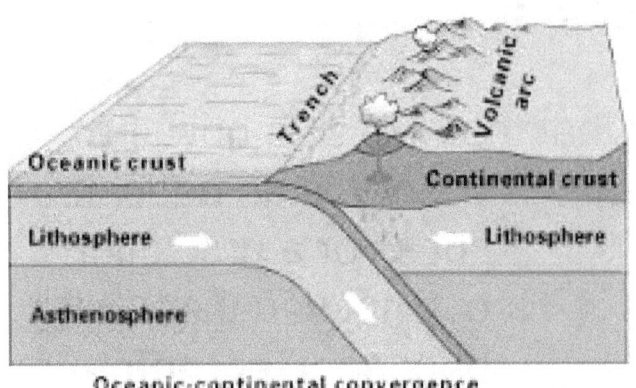
Oceanic-continental convergence

Eurasian plate, Indian plate, Australian plate, and Antarctic plate. all these plates have different sizes. They also contain oceanic and continental crust. The **oceanic crust** is the crust that is located underneath the oceans. The **continental crust** is the crust makes up the continents. The oceanic crust is denser than the continental crust. This because the oceanic crust is made up of elements and materials that are more dense than what makes up the continental crust.

The tectonic plates fit together like puzzle pieces. This is because the tectonic plates were all connected into one, single landmass called "**Pangaea**". Over many millions of years, this landmass broke up and was pushed away from each other. This occured through the action of sea floor spreading.

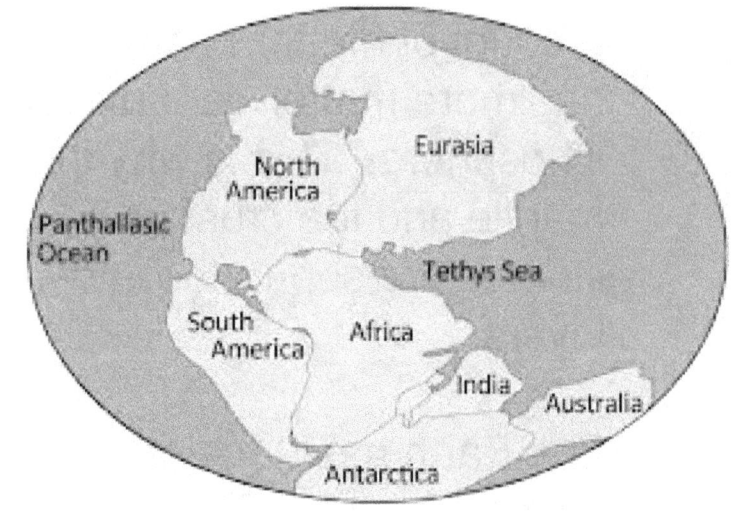

Fractures or breaks in the tectonic plate can cause magma to rise in these locations and fill in these breaks. The magma cools into rock which pushes away the older rock surrounding it. As the older rock is pushed away, continental drift occurs.

Tectonic plates also have different thicknesses. The sea floor or **oceanic crust** is thinner than the continental crust. The **continental crust** can fold and build up over time into plateaus and mountain ranges. The thinnest part of the Pacific plate is located at the mid-ocean ridge. The thickest part is located next to a **plate boundary** or the place where two plates meet. This plate boundary is known as the San Andreas fault. The city of Los Angeles is located on the Pacific plate where this plate and the North American plate meet. These two plates have a similar thicknesses and densities. Because of this, they form a **transform boundary** where one plate slides past the other. The Pacific Plate, at this time, is sliding to the north of the North American plate.

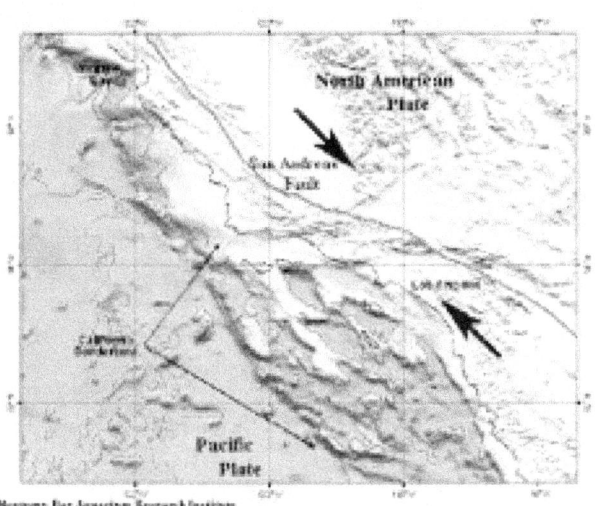
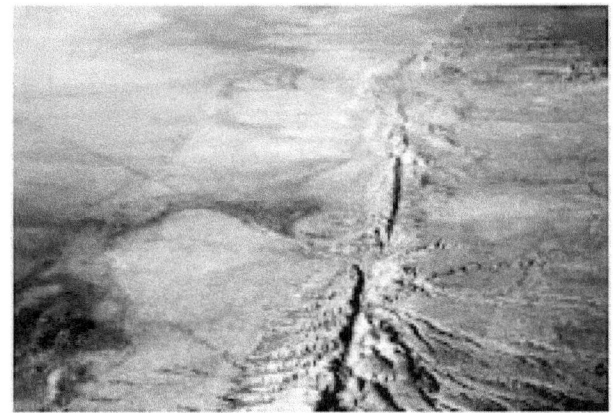

Knowledge and Comprehension:

Plate Tectonics:

Theory:

mantle:

density:

oceanic crust:

continental crust:

Pangaea:

Plate Boundary:

Transform Boundary:

1. What is plate tectonics?

2. Identify the Earth's 9 major tectonic plates.

Application, Analysis, Evaluation, Synthesis

3. What is the difference between continental and oceanic crust? Which one is more dense? Why?

4. Explain why the continents fit together like a jigsaw puzzle.

5. Explain what a transform boundary is. What is a good example of this? Explain why this is important to know.

Tectonic Plate Boundaries

Tectonic plate boundaries are places on the Earth where tectonic plates meet. These boundaries are determined by the presence of volcanoes, mountains, trenches, ridges, plateaus and earthquakes. There are 3 different types of tectonic boundaries: Transform boundaries, convergent boundaries, and divergent boundaries.

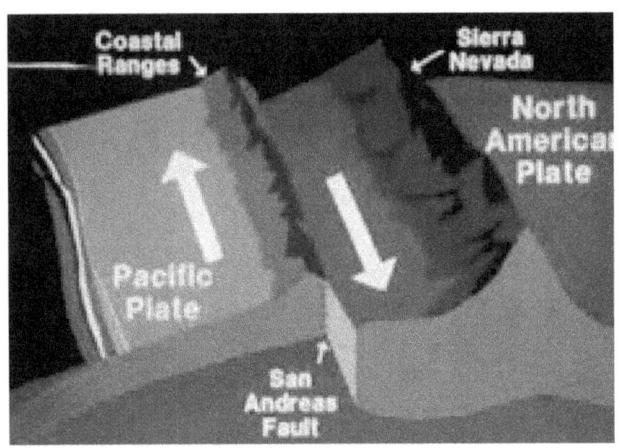

Transform boundaries are places on the Earth's surface where tectonic plates meet and slide past one another. Plates can slide back and forth or up and down. One example of this is the San Andreas fault. This is the location where the Pacific plate and the North American plate slide past one another. The fault occurs through the oceanic crust and the continental crust. The Pacific plate is moving north in direction.

Convergent boundaries are places on the Earth where tectonic plates collide together. Collisions such as this can cause mountain ranges to form. One example of this is the African plate colliding with the Eurasian plate. As a result of this, the Himalayas, the world's tallest mountain range was formed. Subduction can also result from the collision of two tectonic plates. **Subduction**, the movement of a plate underneath another plate, can occur. The more dense plate subducts or moves underneath the less dense plate. Subduction causes

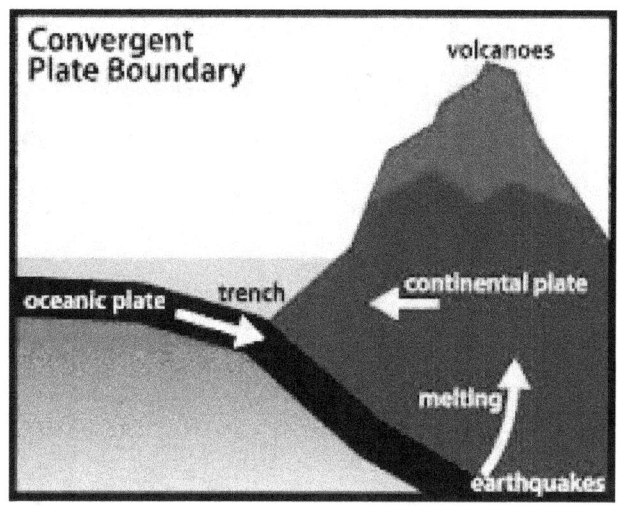

earthquakes to occur, land volcanoes, volcanic islands (**volcanic arc**) and mountain ranges to form parallel to the boundary. One real and frightening example of this process was the earthquake near Fukishima, Japan. This earthquake registered a 9.0 on the Richter scale. This earthquake was caused by the Pacific plate subducting or sinking below the Eurasian plate. It has been reported by scientists that this event caused the angle of the Earth's axis to shift or change.

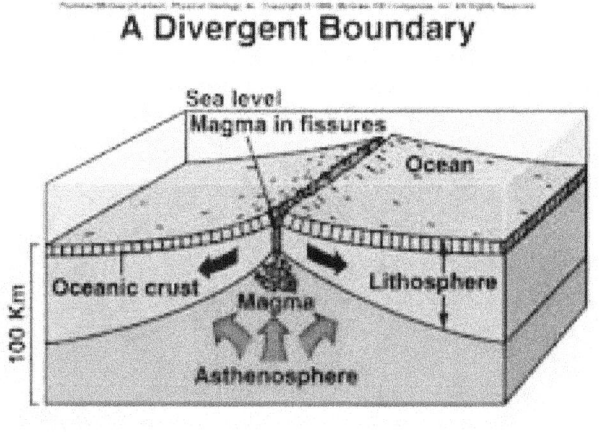

Divergent boundaries are places on the Earth where tectonic plates move away or separate from each other. These boundaries mostly occur on the ocean floor at mid-ocean ridges. **Mid-ocean ridges** are volcanic mountain chains that are formed parallel to the boundary from magma that rises up through breaks in the ocean floor or crust. As this magma cools down and hardens, it pushes the older rock away from the boundary.

Knowledge and Comprehension:

Tectonic Plate Boundaries:

Transform Boundary:

Convergent Boundary:

Divergent Boundary:

Subduction:

Volcanic Arc:

Mid-Ocean Ridge:

1. Identify the three types of tectonic plate boundaries.

2. What is subduction?

3. How are convergent boundaries and subduction related?

Application, Analysis, Evaluation, Synthesis

4. Explain what occurs at a mid-ocean ridge.

5. Explain how mountain ranges form.

6. Identify what the major differences are between the three types of tectonic plate boundaries.

Continental Drift

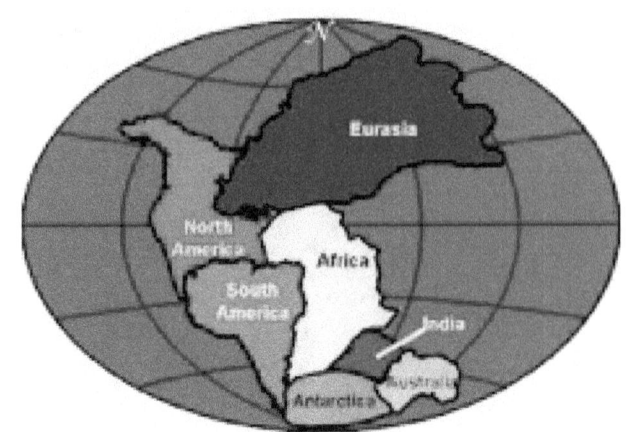

Continental drift is a theory, an educated guess, that was proposed by Alfred Wegener in the early 1800's. He stated that a large landmass broke up into smaller pieces. These smaller pieces are the 7 continents that we know today as Antarctica, North America, South America, Africa, Europe, Asia and Australia. The **continents** are landmasses that are part of the Earth's crust. This large landmass was given the Greek name "**Pangaea**." "Pan"- means "all" and "gaea" means "Earth". Gaea was also the name of the Greek goddess of the Earth in ancient times.

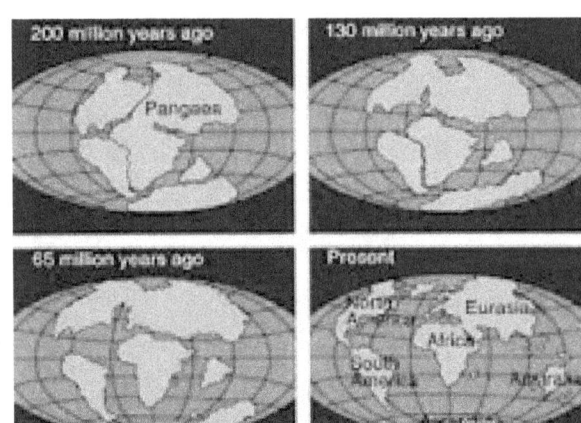

Pangaea existed about 245 million years ago. 135 million years ago, it broke up into 2 large pieces called **Laurasia** and **Gondwana.** About 65 million years ago, Laurasia broke up into the continents of North America, Europe and Asia. Gondwana broke up into the continents of Africa and South America. Through out this time, each of the pieces that broke off from Pangaea have drifted away from each other, while some of them have collided with each other forming mid-ocean ridges, ocean trenches, volcanoes and mountain ranges.

What evidence do we have that continental drift occurred?

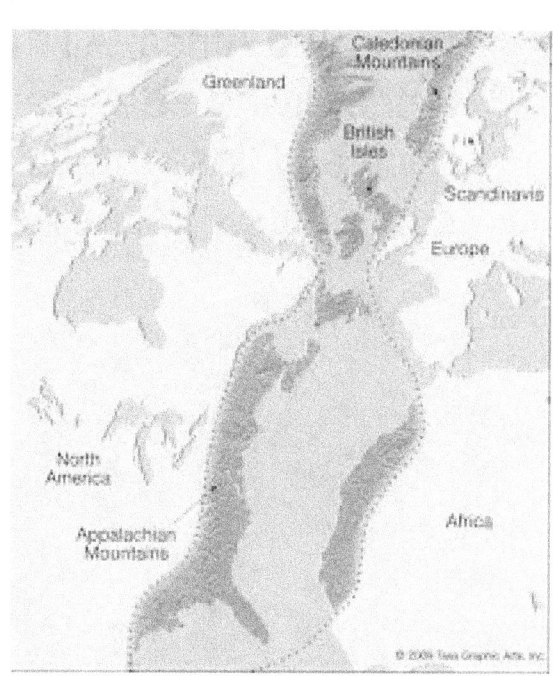

There is several types of evidence confirms that continental drift occurred. Mountain ranges, like the ones in Greenland and the Eastern United States match the mountain ranges in the British Isles, the Western Europe and Northwestern Africa. They have similar types of rocks within them. The outline and shapes of their coastlines fit together like puzzle pieces.

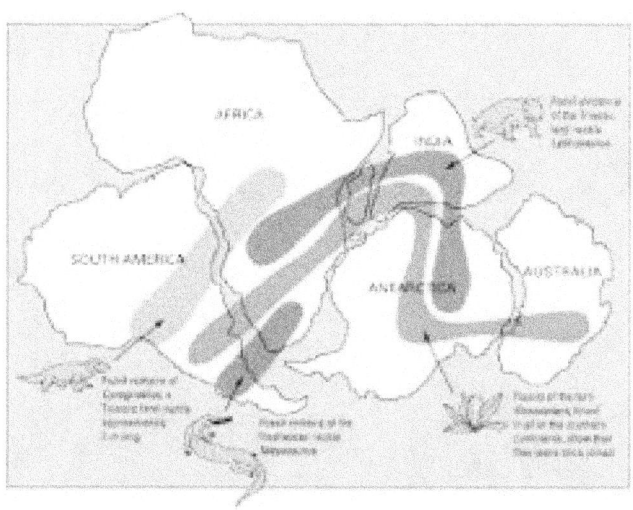

Fossils of ancient organisms have been found on different continents around the world. These organisms are known as Mesosaurus, Cynognathus, and Lystosaurus. Mesosaurus was a dinosaur, a freshwater reptile, that lived about 56 million years ago. It's fossils have been found in Brazil and Africa. Cynognathus, and Lystosaurus are both land reptiles that lived during the Triassic period. Cynognathus fossils have been found in Argentina, South America and Africa. Lystosaurus fossils have been found in Antartica, India and Africa. All these species are now **extinct** or no longer living. Several fossils of different plant species such as Glossopteris were also found in all of the southern continents such as Antarctica, South America and Africa.

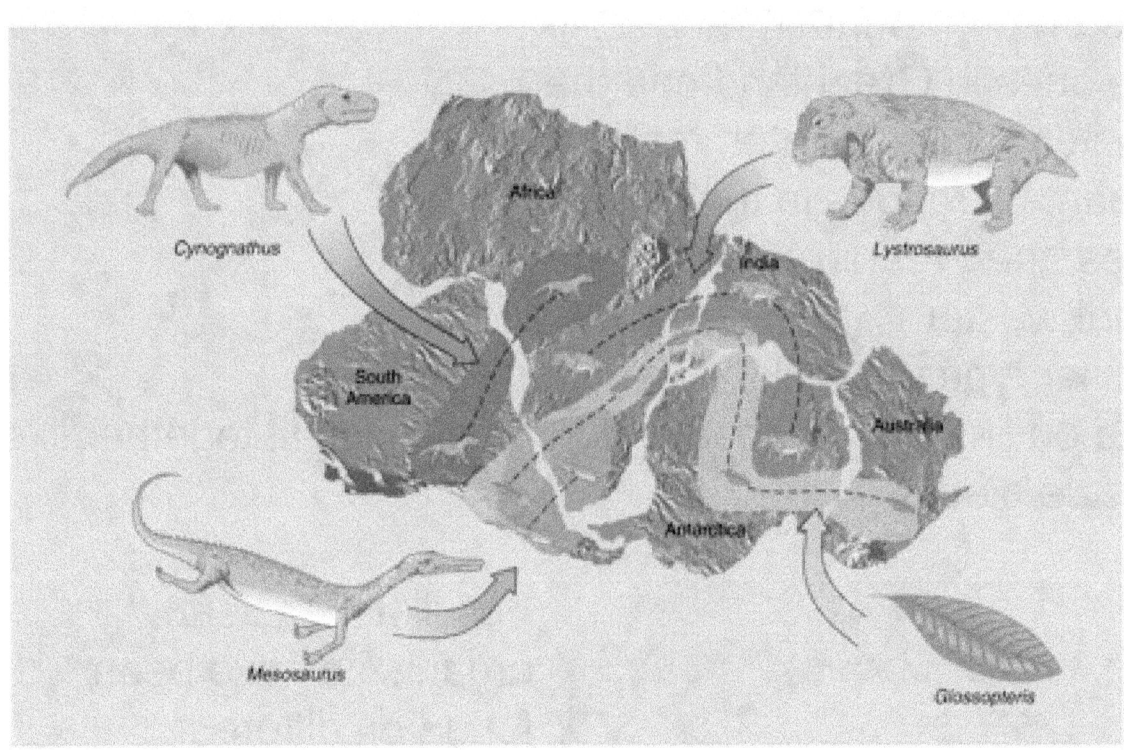

Knowledge and Comprehension:

Continental Drift:

Continents:

Pangaea:

Laurasia:

Gondwana:

Extinct:

Mesosaurus:

1. Describe what Pangaea is.

2. What does the word "Pangaea" mean?

Application, Analysis, Evaluation, Synthesis

3. Explain the theory of continental drift.

4. Show what happened to Pangaea during the following time periods by drawing pictures and labeling them. Also write down and explain what is happening in words.

250 million years ago:

135 million years ago:

65 million years ago:

5. What are the 3 different types of evidence that support the theory of continental drift? How do they confirm that continental drift actually occurred.

6. Color the following diagram. Explain what this diagram is trying to represent.

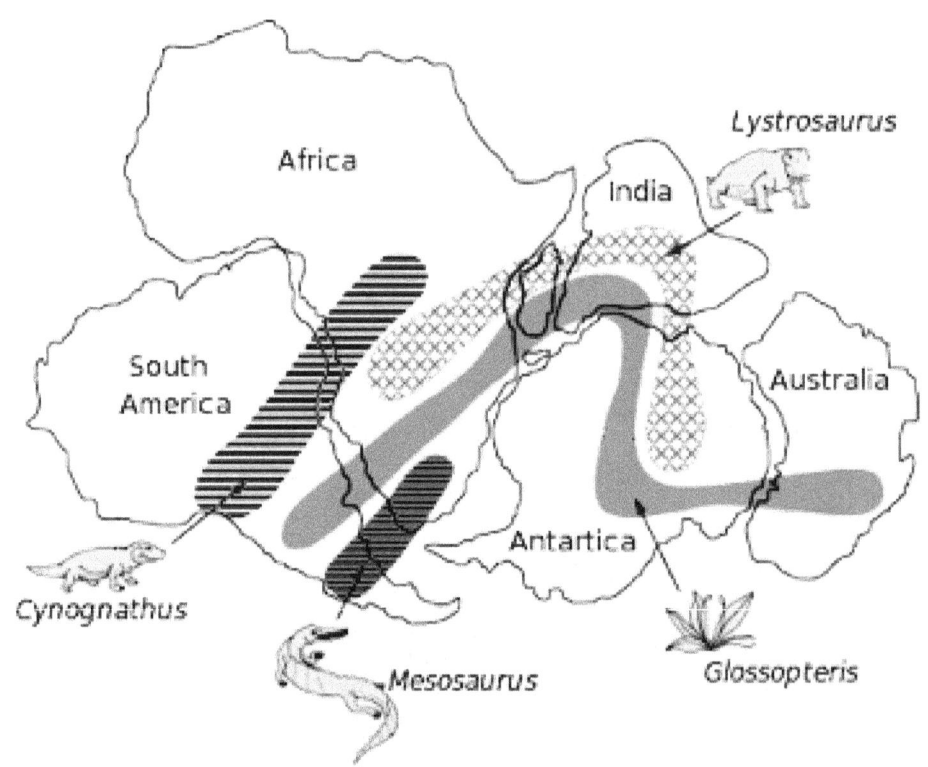

Mid-Ocean Ridges and Sea Floor Spreading

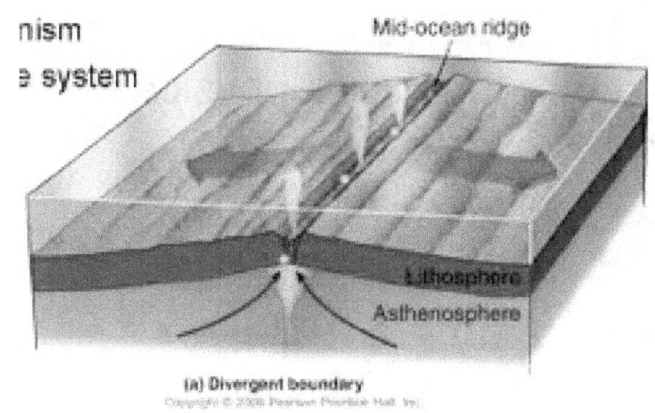

All around the world there are **mid-ocean ridges** that exist on the bottom of the sea floor. They are chains of mountains that are formed from the rising of magma from the Earth's mantle onto the Earth's crust. The **magma**, or hot, molten rock, rises through fractures or breaks in the sea floor. The magma forms new **sediment** or rock as it cools. As the rock builds up from constantly depositing new rock, mountains or ridges are formed.

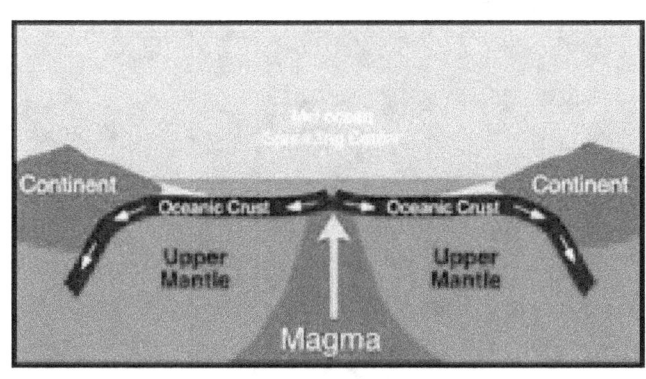

As the **tectonic plates** or portions of the Earth's crust move away from one another, the older rock moves away or pulls apart from the mid-ocean ridge. Through this action, sea floor spreads apart. This process is called **sea floor spreading**. This is how new sea floor is formed.

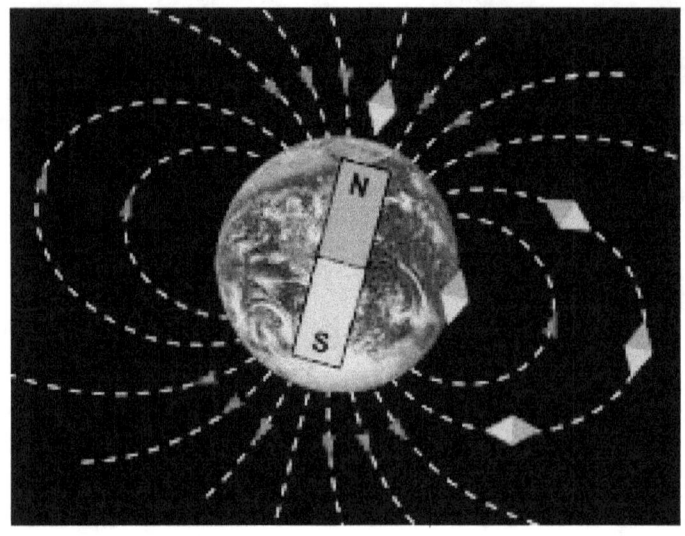

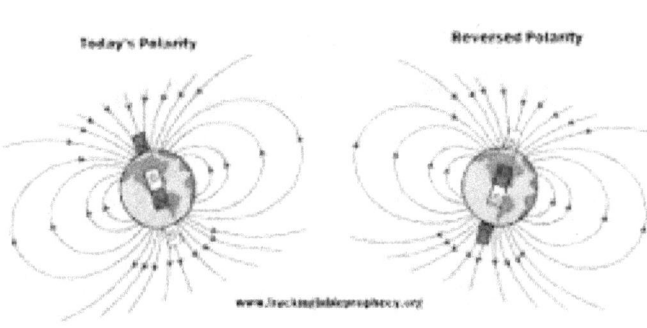

As the sea floor spreads, it brings a record along with it. This "record" is magnetic. It physically records the magnetic reversals that occur periodically on Earth. A **magnetic reversal** is a change in the direction of Earth's magnetic field at the north and south poles. Magma, because it contains iron in it, is affected by the magnetic field of the Earth. The Earth's **magnetic field** is like a shield of electromagnetic energy that is created by the iron within the core or center of the Earth.

When the magnetic field of the Earth changes direction, the magnetic fields of the pieces of iron inside the magma align or line up with the magnetic field of the Earth. The magma, as it cools down, will become a permanent record of the

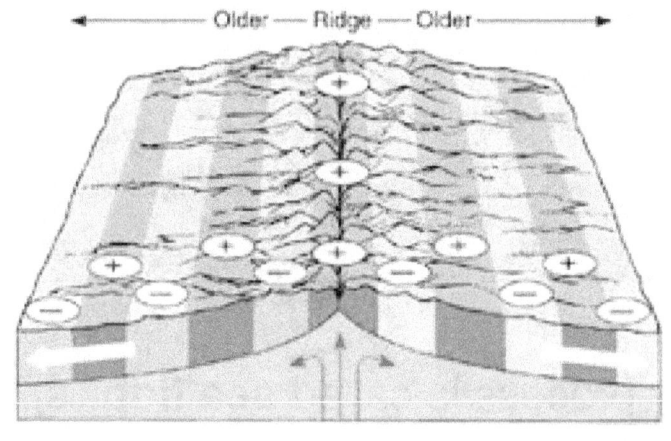

direction of the magnetic field at different points in time. This record also serves as evidence that the continents are moving. Continental drift, the process by which the continents move from one location to another location on Earth, is caused by the spreading of the sea floor and the subduction of the tectonic plates.

Knowledge and Comprehension:

Mid-Ocean Ridge:

Magma:

Sediment:

Tectonic Plates:

Sea Floor Spreading:

Magnetic Reversal:

Magnetic Field:

1. What is the Mid-Ocean Ridge?

2. What is sea floor spreading?

Application, Analysis, Evaluation, Synthesis

3. Explain how mid-ocean ridges are created. Also, draw a picture and label it.

4. Explain how a magnetic record is created as the sea floor spreads.

5. Explain how sea floor spreading is evidence that supports continental drift.

Subduction

Subduction is the process by which a **tectonic plate**, or a piece of the Earth's crust, subducts or moves under another plate. The reason why subduction occurs is because one plate is more dense than the other. Oceanic crust, because it is made up of dense material and elements, is more dense than continental crust. Because of this difference in density, the oceanic crust sinks under the continental crust. **Density** is measured by the amount of matter within a certain space or volume.

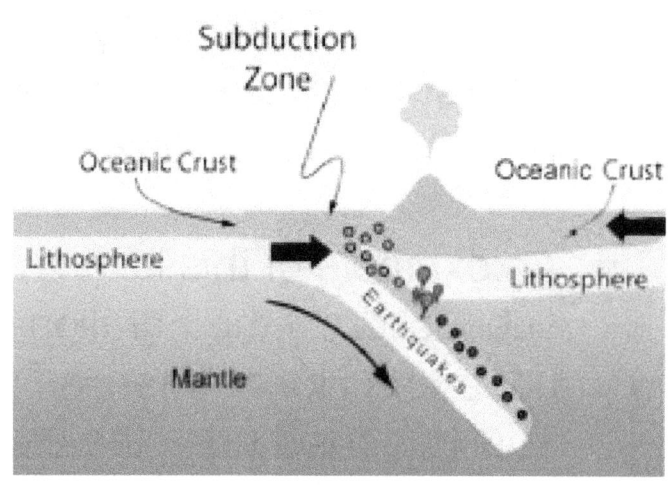

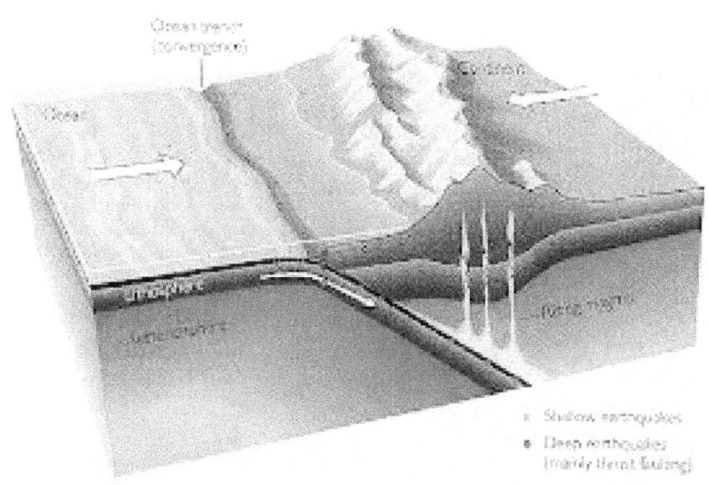

As a plate **subducts** or sinks underneath another plate, friction is created. Any time friction is created, heat is given off. This heat is transferred and the rock within the tectonic plate begins to melt and become a liquid. Water that was trapped inside the rock is given off as steam or water vapor. The melted rock then sinks, by gravity and mixes into the mantle of the Earth. In this way, older sediment or rock existing on the Earth is destroyed to make way for the creation of new sediment or rock. New rock is continually made at the mid-ocean ridges. The **mid-ocean ridge** is a break in the sea floor or the crust where magma is allowed to rise up.

An example of this is the subduction of the Farallon plate under the North American plate. This has resulted in the creation of the Sierra Nevada mountain range and the **uplift** or rising of the Colorado Plateau, or the four corners region of the United States. The four corners region includes the states of Nevada, Colorado, Utah, and Arizona. A **plateau** is a region of land that has been uplifted or risen by excess magma that has built up underneath it. This magma originated or came from the melting of the ancient Farallon plate during its subduction. As the plate is pulled down into the mantle by gravity, it creates friction. This friction creates a lot of heat which raises the rock to a very high temperature. This high temperature melts the rock into

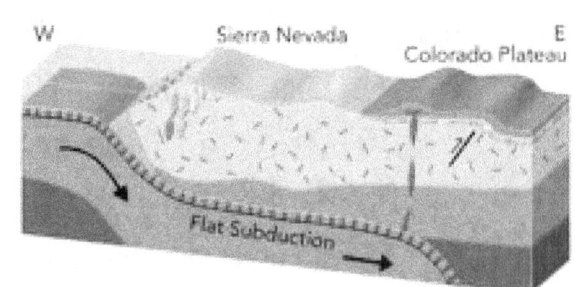

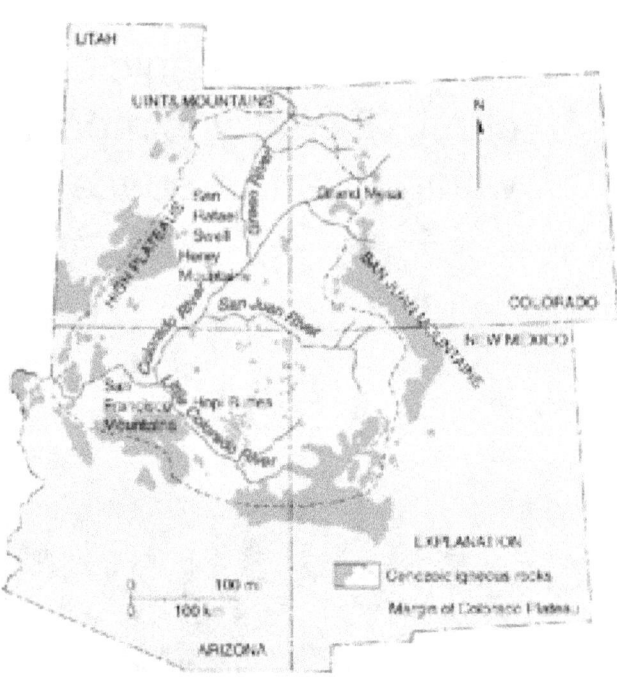

magma which mixes into the Earth's **mantle** or layer of magma. This process is called **sea floor spreading**. This process pushes and moves land away from the mid ocean ridges.

Together with sea floor spreading, the Earth continually creates new land and destroys old land. New land is created at the mid ocean ridges through sea floor spreading. While this is occurring, oceanic plates are continually **subducting** or sinking underneath continental plates and being destroyed. These two processes balance each other by allowing new land to always be created and destroying land that is older.

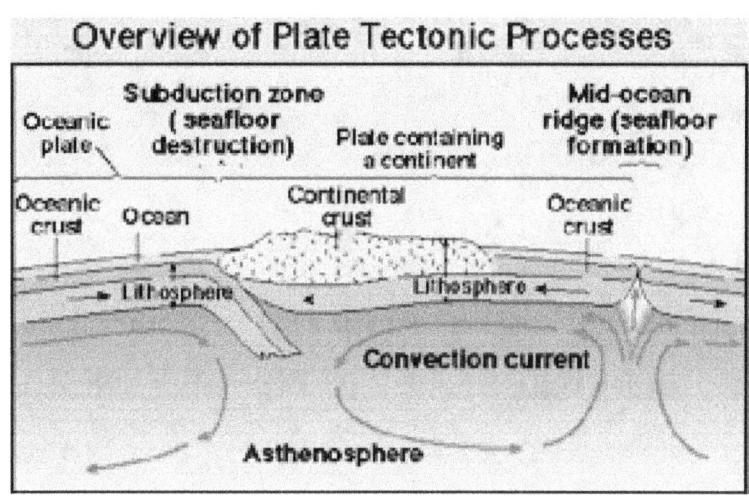

Knowledge and Comprehension:

Subduction:

Tectonic Plate:

Density:

Subduct:

Mid-ocean Ridge:

Uplift:

Plateau:

Mantle:

Sea Floor:

1. Describe the process of subduction.

2. In order for subduction to work, identify the other process that must occur.

3. Which plates are more dense: oceanic or continental plates. Why?

Application, Analysis, Evaluation, Synthesis

4. Explain what happens to a plate when it subducts under another plate. Also, draw a picture and label it.

5. Explain how the Sierra Nevada Mountain range was formed.

6. What is a plateau? Explain how the Colorado plateau was created.

What is a Volcano?

A **volcano** is a vent or opening in the earths surface through which magma and gases that are under extreme temperature and pressure, escape from inside the Earth. **Magma** is rock that has been melted at high temperatures and pressure within the mantle layer of the Earth.

Sometimes, this mantle will punch holes into the crust creating a magma chamber or a reservoir of magma underneath the earth's surface. When the magma is released during a volcanic eruption onto the surface of the Earth, it is called **lava**.

The heat generated by the magma causes the pressure within the volcano to build up. The solid rock surrounding the magma chamber melts into liquid rock, or magma, at lower temperatures as the pressure increases. As the magma chamber grows in

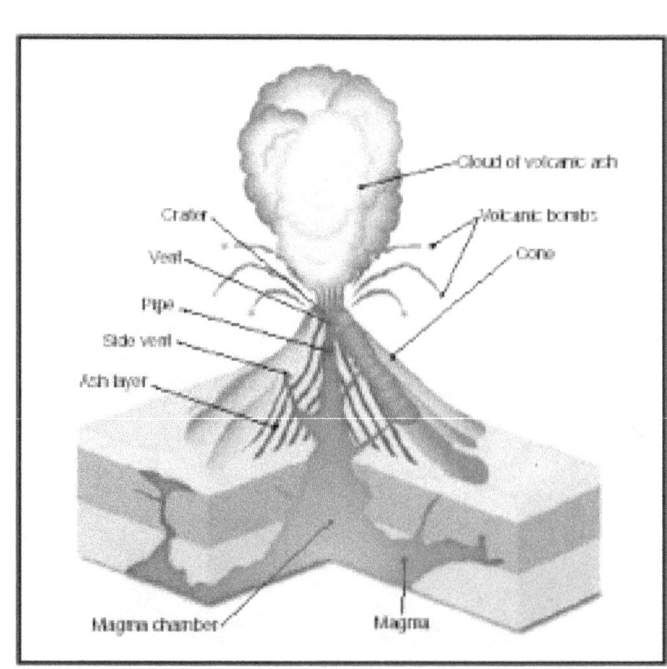

volume, the land above it begins to rise, and the land around it begins to expand and form fissures or cracks. **Seismic activity** in the form of frequent earthquakes can be felt for miles.

Volcanic eruptions could be mild to extreme depending on the type of volcano. Some volcanoes erupt with tremendous force and power. This type of eruption releases ash, lava, and gases into the atmosphere. Sometimes deadly pyroclastic flows occur. **Pyroclastic flows** are made up of ash and gas that has been heated up to 900 degrees Fahrenheit or more. Towns and villages in the ancient world, such as Pompeii and Herculaneum have been decimated by pyroclastic material. They are dangerous because they have been known to disintegrate any type of living and non living thing that is in its path.

Focus Questions:

1. Describe what a volcano is.

2. Describe what magma is.

3. What is the difference between magma and lava.

4. Explain why a volcano erupts.

5. Why are pyroclastic flows dangerous?

Ancient Volcano: Mt. Vesuvius

Mt. Vesuvius is an active volcano that is located in the province of present day Naples, Italy. It is a stratovolcano that is located on the coast, a very short distance from the seashore. It is part of the Campanian volcanic arc.

In ancient times, Mt. Vesuvius erupted and buried the ancient Roman cities of Pompeii and Herculaneum in 79 AD. Villas were destroyed and buried underneath 4-6 meters of ash and pumice. **Pumice** is a volcanic rock that is made up of rough, volcanic glass.

The towns of Pompeii and Herculaneum were rediscovered in 1599. Walls

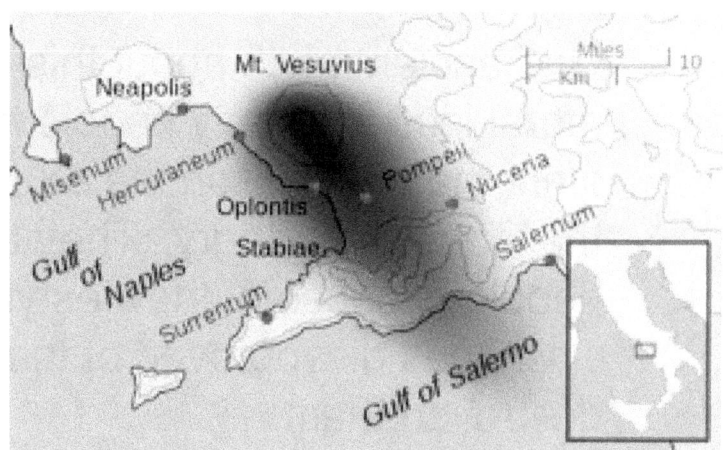

of the ancient cities and inscriptions were found while work on the construction of a water channel was underway. Guiseppe Fiorelli, in 1863, made a gruesome discovery. He found voids or empty space in the ash layer that contained human remains. He noticed that these spaces were left behind were left behind by humans. He used the technique of injecting plaster into these voids to recreate the humans that were once there. It became very clear to him that the victims of eruptions died most likely from inhaling hot ash and pumice and suffocating. Scientists estimate that these people were exposed to to heat at the temperature of 250 degrees Celcius, at a distance of 10 kilometers away from the vent of the volcano.

One ancient myth about this event was fashioned into a moral lesson. Many people that lived away from the cities state that the reason the eruption occurred was because the Gods were angry. They became enraged at the people because of the lifestyles that they were leading. The Gods disapproved of their "barbaric" behavior and destroyed them.

Source:

http://en.wikipedia.org/wiki/Pompeii

Focus Questions:

1. Where is Mt. Vesuvius located?

2. When did Mt. Vesuvius erupt?

3. Which cities were destroyed by the eruption?

4. What affected the people that were living in the city during the eruption? Did they survive? Why or why not?

5. How did this event evolve into a myth? Explain.

Yellowstone Supervolcano: Mantle Plume is 2.5 x Bigger Than Previously Thought

Yellowstone National Park is a geothermal active region of the United States. This activity, known for its creation of heat from deep underneath the crust of the Earth, is caused by a huge **mantle plume** that sits right underneath the park.   A mantle plume is a reservoir or chamber of hot magma that punctures the crust of the Earth and rises up underneath the surface of the crust. Geologists call these locations on the Earth, "**hot spots**."

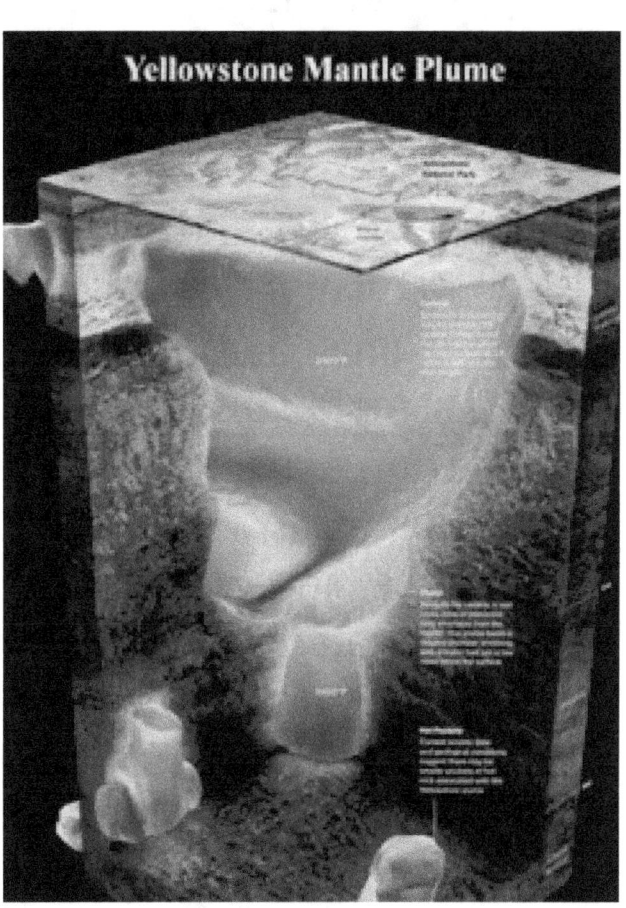

Yellowstone is also well known as a **super volcano**. A duper volcano is a massive volcano that is capable of erupting at anytime and causing damage that is widespread when compared to other volcanoes. The last time the Yellowstone

super volcano erupted was over 600,000 years ago. Evidence of previous **calderas**, or sites where a volcano has erupted leaving a depression in the Earth's crust, from this volcano have been identified. The magma plume of the Yellowstone magma chamber is estimated to be 80 kilometers long, and 20 kilometers wide. This 2.5 times bigger than originally estimated according to Robert Smith, a geophysicist from the University of Utah, in Salt Lake City. This may indicate that the magma chamber has expanded when compared to data that was previously analyzed. If the chamber did become larger, it could have melted the rock around the magma chamber, through increased pressure and heat, widening its volume.

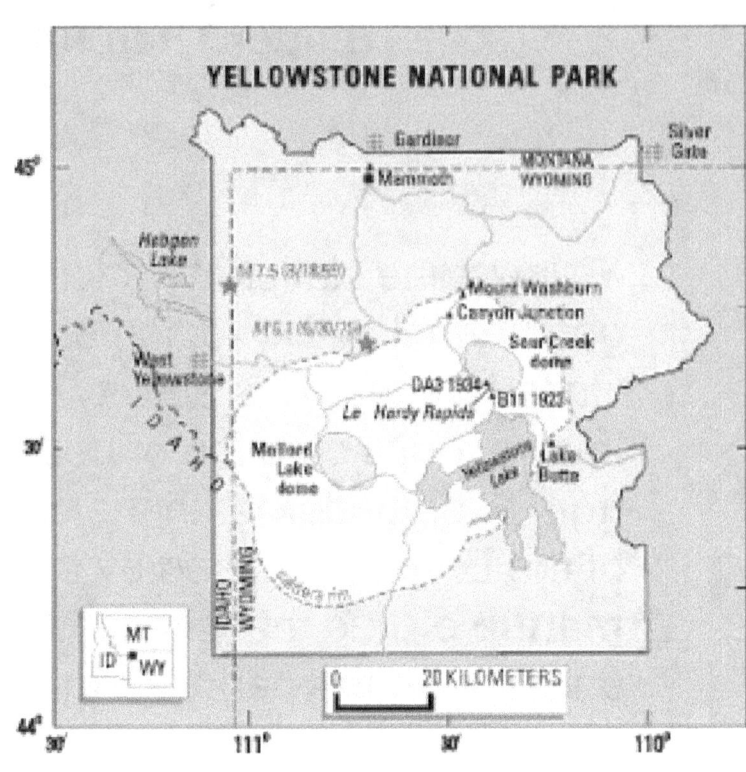

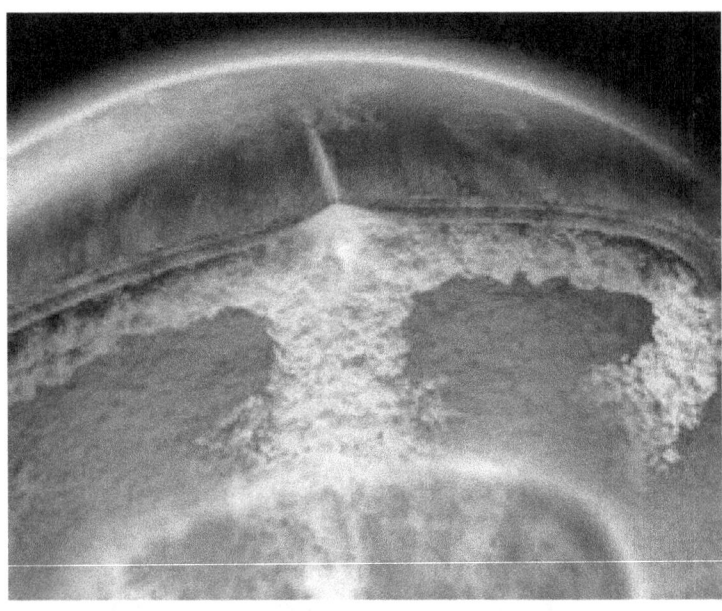

Scientists have noticed that the chamber has been rising at an alarming rate since 2004. The last three years, records have indicated that the floor has been rising at a rate of 3 inches per year, the fastest rate side record keeping began in 1923. The region has also experience many swarms of earthquakes in the past 3 years.

Source:

http://www.earthmountainview.com/yellowstone/yellowstone.htm

Focus Questions:

1. What is a mantle plume?

2. What sits beneath Yellowstone National Park?

3. Is Yellowstone a super volcano? Explain.

4. What evidence suggests that the size of the Yellowstone magma chamber has been increasing?

5. What do you predict will happen if this super volcano erupts?

What are Earthquakes?

An **earthquake** is a sudden release of energy in the crust of the Earth. This seismic energy travels through the crust in the form of waves. **Energy** is defined as the ability of an object to do work over a specific distance. This can be described in the mathematical formula, work = force x distance. The strength of the earthquake depends on where the epicenter is located and how much energy is released.

Earthquakes are most commonly experienced in places where the tectonic plates move and slide past each other or when one plate **subducts** or is pulled underneath the other by the force of gravity. This occurs around the Pacific Rim

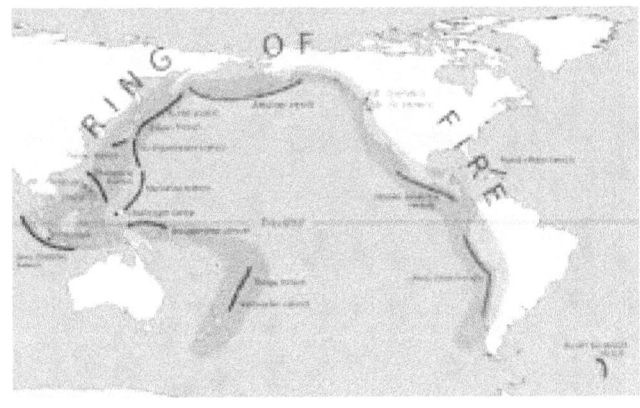

and is called the the "**Ring of Fire**. Earthquakes also occur at faults where there is a break in the Earth's crust. One famous fault, the **San Andreas fault** in California, separates the North American plate from the Pacific plate. The two plates are sliding past one another in the north and south directions. The city of Los Angeles is located on the Pacific Plate and is moving north towards San Francisco.

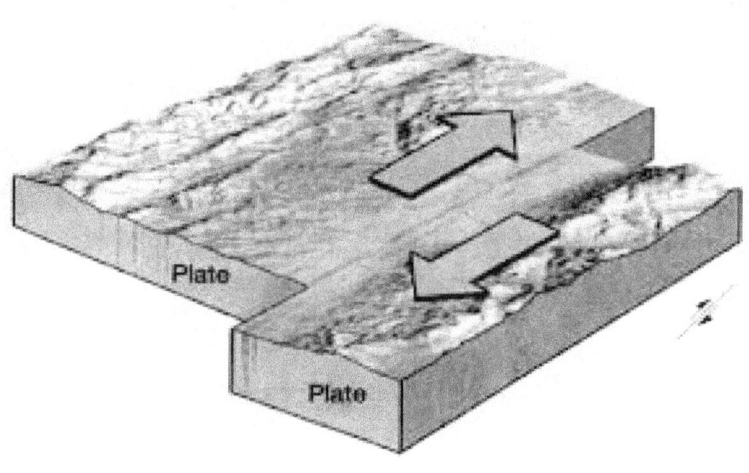

The energy that is released during an earthquake could be the result of the building up of force and pressure in a fault zone or between two large slabs of the Earth's crust during a period of time. Sometimes two slabs of crust slide past one another horizontally and others slide past each other vertically. Some times the rock in the crust cause resistance, which is a force, and inhibits the movement of the two slabs of crust. The rock in the crust, during this time, can be deformed and stretched. When the 2 slabs of crust finally move from the location where they were locked up, the pressure and force is released and given off as energy.

Earthquakes cause shaking of the ground and landslides to occur. The intensity of the shaking depends on how much energy is released during the seismic event and the type of medium the energy travels through, such as rock, sediment,

sand and water. They could also displace the surface of the Earth in some locations or cause new breaks or **fissures** in the crust. The intense shaking can cause buildings to collapse and fires to break out.  If an earthquake occurs off shore, movement of the sea floor could cause a tsunami or a huge wave of sea water to occur.

Focus Questions:

1. What is an earthquake?

2. What causes an earthquake?

3. Explain what happens to the energy that is released in an earthquake. How is this energy created?

4. What is a fault zone? Describe what can occur here.

5. What type of damage can an earthquake cause?

6. How can this damage be minimized?

What is a Tsunami?

A **tsunami** is a large wave that is generated by an earthquake or movement of the sea floor. The word tsunami is a Japanese word that means "harbor waves". Earthquakes as well as volcanic eruptions, underwater explosions, glacier calvings, landslides and meteorite strikes can cause tsunamis to occur.

A tsunami that originates in the ocean is generated by the vertical rising of the sea floor at the epicenter of an earthquake. As the sea floor rises, the water above it rises as well. This displaces a huge volume of water that becomes displaced and begins to move away from the epicenter, The energy that is released from the movement of the seafloor is transferred to the surrounding water and causes an increase in its speed.

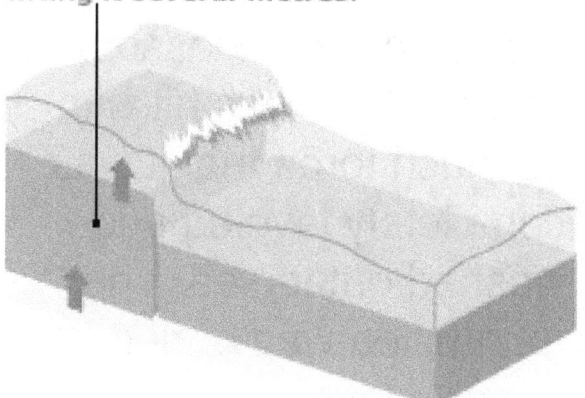
Earthquake vertically jolts seabed, lifting it several metres.

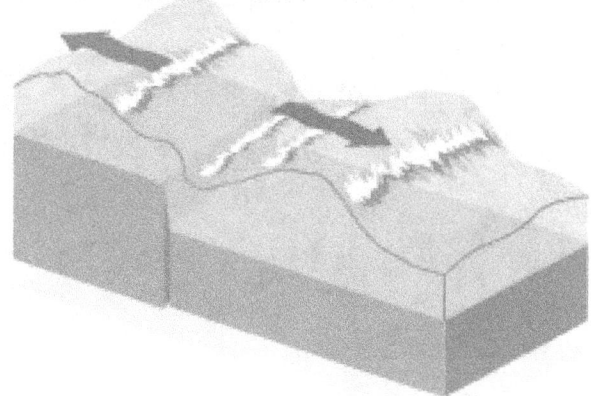
Large waves begin moving through the ocean, away from the epicentre.

Tsunamis traveling at a high speed with a tremendous amount of energy can strike the coastlines of many continents and islands with an incredible amount of force. The **force** is equivalent to the acceleration of the incoming wave multiplied by the distance it travels. Depending on the location of the epicenter, it can take a few minutes or hours to travel. The closer the epicenter is to land, the more damage it will likely make.

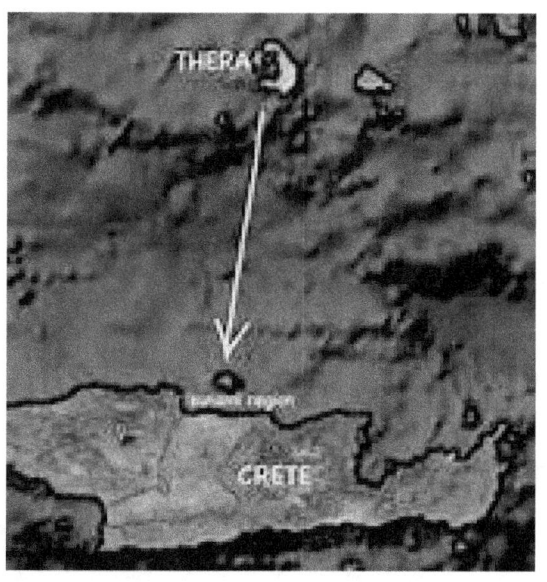

Tsunamis have been reported to destroy entire villages, towns, and civilizations. The Minoan civilization on the island of Crete was destroyed in 1600 BC by the eruption on a volcano on the neighboring island of Thera or Santorini. The tsunami that was

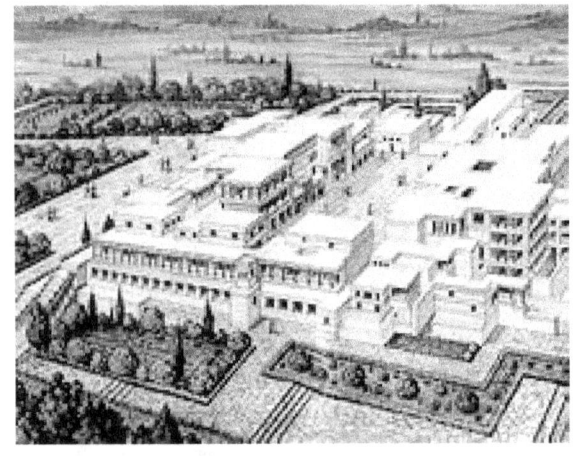

generated by the volcanic explosion destroyed the coastal city of Knossos. Its great palace was hit with so much force that it was completely leveled and washed out to sea. The foundation was buried under a layer of sand sand. Many of the people that lived on the island abandoned their homeland for what is now the Greece, Turkey, and Syria. Earthquakes before the eruption served as a warning to leave the island before the catastrophe occurred. After the tsunami devastated the island, there "was nothing to come home to."

Focus Questions:

1. What is a tsunami?

2. What can cause a tsunami to occur?

3. Describe how a tsunami originates from an earthquake that occurs on the sea floor.

4. Why is a tsunami dangerous?

5. Describe what happened to the island of Crete after the volcanic explosion on the island of Thera or Santorini. What happened to the Minoan civilization?